Institute of Terrestrial Ecology
Natural Environment Research Council

Distribution of Freshwaters in Great Britain

Ian Smith and Alex Lyle

Institute of Terrestrial Ecology
Edinburgh

Published in 1979 by
Institute of Terrestrial Ecology
68 Hills Road, Cambridge CB2 1LA

ISBN 0 904282 25 2

Cover: Gairney Water at Powmill (Photo: K. East)
Enquiries to I R Smith
Institute of Terrestrial Ecology
78 Craighall Road
Edinburgh
EH6 4RQ
(031-552 5596)

The Institute of Terrestrial Ecology (ITE) was established in 1973, from the former Nature Conservancy's research stations and staff, joined later by the Institute of Tree Biology and the Culture Centre of Algae and Protozoa. ITE contributes to and draws upon the collective knowledge of the fourteen sister institutes which make up the *Natural Environment Research Council*, spanning all the environmental sciences.

The Institute studies the factors determining the structure, composition and processes of land and freshwater systems, and of individual plant and animal species. It is developing a sounder scientific basis for predicting and modelling environmental trends arising from natural or man-made change. The results of this research are available to those responsible for the protection, management and wise use of our natural resources.

Nearly half of ITE's work is research commissioned by customers, such as the Nature Conservancy Council who require information for wildlife conservation, the Forestry Commission and the Department of the Environment. The remainder is fundamental research supported by NERC.

ITE's expertise is widely used by international organisations in overseas projects and programmes of research.

Introduction

The object of this study is to estimate from the examination of maps the numbers, distribution and size of the lakes, reservoirs and rivers in Great Britain. The study was undertaken to provide a framework for the systematic analysis of data on freshwater ecosystems. A coded checklist of animals occurring in freshwater in the British Isles has already been published (Maitland, 1977).

The detailed procedures used in the map analyses and checks on the effects of map scale on which waters are included are considered separately for standing and running waters. The overall extent and distribution of lakes, reservoirs and rivers are then reviewed in the third chapter. Almost all statistical data relating to water in Britain are tabulated in terms of Hydrometric Areas (Water Resources Board, 1974) and this practice has been followed here (see Figure 1).

We would like to thank Miss Zana Juppenplatz and Mr Mervin Hutchinson for their valuable assistance with the map searches.

1 The lake and reservoir survey

The basis of the survey is a count of all standing waters marked on the 1:250000 (quarter inch to one mile) maps of Britain. As far as is known, there is no precise criterion for including any water body on such maps. The identified population, therefore, is strictly a large sample selected on a rather uncertain basis. A complete population of waters above a certain area, say 1 km^2, can, however, be identified, the uncertainty applying only to the smaller lakes and reservoirs. The adequacy of coverage using maps of this scale is discussed on p 6.

INFORMATION OBTAINED DIRECTLY FROM MAPS

Master maps were prepared, from which the Hydrometric Area, National Grid reference to the centre of gravity of the water body, name (if marked) and altitude interval as indicated by the contour layer colouring were obtained directly. The water surface areas were measured using a grid overlay and then grouped into classes, the class intervals used being logarithmic, viz, less than .25 km^2, between .25 and .50 km^2, between .50 and 1 km^2, between 1 and 2 km^2 and more than 2 km^2. The information was tabulated for every water body in files sub-divided by Hydrometric Area. For those Hydrometric Areas which cross national boundaries, each water body is assigned to the appropriate country so that national totals can be derived.

It was noticed that, in several cases, the information presented on the 1:250000 maps differed from that on larger scale maps. No corrections were made to the basic data for any observed over-simplifications in order to retain their consistency. Seventeen map sheets were used in the survey and their date of correction is not always the same. This means that there is some uncertainty about the number of reservoirs included but, as far as is known, the number of water bodies is correct as at the 31 December 1968.

The actual areas of waters estimated from the grid overlay to be more than 2 km^2 were determined in a number of ways and noted in the files. Because the sources of information are different, it is easier to consider Scotland separately from England and Wales.

In Scotland, only a very few of the natural lakes at this size were not surveyed by Murray and Pullar (1910). The areas of artificial reservoirs and natural lakes affected by the operations of the North of Scotland Hydro-Electric Board were supplied by the NSHEB (Pers. Comm.). Information on waters used for public water supply were obtained from a publication by the British Waterworks Association (1972). Fourteen waters remained, mainly reservoirs constructed since Murray and Pullar's survey, and the areas of these were measured directly with a planimeter using maps at a scale of 1:63360 (one inch to one mile).

In England and Wales, there are only 26 waters of this size and most of the natural lakes are in the Lake District for which data could be obtained from Mill (1890). The remaining areas were either measured directly in the same way as those in Scotland or, for several recently constructed reservoirs, individual descriptions were consulted.

The results of the count are given in Appendix 1. The total number of water bodies is 5505 of which more than two thirds are in Scotland.

THE ADEQUACY OF THE COUNT

The testing of the adequacy of the original count is complicated by recent changes in the form of large scale maps published in the British Isles, the traditional maps at a scale of 1:63360 (1 inch to one mile) having been replaced by a new series at a scale of 1:50000. The first edition of the series is derived directly from the older form of map so that the numbers of standing waters marked on them should not have changed, but differences in numbers are possible on the second edition. Since only a quarter of the 1:50000 maps so far published are of the revised format, and most of these are in the Highlands and Islands of Scotland, the resolution of the original count has been tested comparing numbers on sample areas on the earlier 1:63360 maps with those on the 1:250000 sheets.

Several colleagues have kindly made available the results of map searches at the larger scale – Mr C. A. Badenoch for the Tweed Basin (Hydrometric Area 21), Mr R. H. Britton for the Shetland Islands and Dr P. S. Maitland for Outer Hebrides. The comparative number of water bodies marked on the two scales of map are shown in Table 1. Since the detailed counts for the first two of these split the numbers of waters into size classes it

Table 1 Comparative counts of lakes and reservoirs on different map scales

Area	No of waters on 1:63360 map	No of waters on 1:250000 map	Percentage
(a) Complete areas			
Tweed	324	38	11.7
Shetland	1550	195	14.4
Outer Hebrides	6038	1094	18.1
(b) 30 km x 30 km grid samples			
1	452	40	8.9
2	70	15	21.4
3	50	8	16.0
4	97	28	28.9
5	185	25	13.5
6	34	5	14.7
7	98	7	7.1
8	52	2	3.8
9	247	2	0.8
10	170	10	5.9
11	766	5	0.7
12	35	1	2.9

is possible to come to some conclusion about the minimum area of water bodies marked on the 1:250000 maps. In the Tweed Basin, the number of waters not counted in the original survey is 286 and this number is exactly equal to the number less than 4 ha. In Shetland, the difference in total numbers between the two counts is 1355 while the number of waters less than 4 ha is 1390. These figures suggest that the count includes all water bodies whose surface area is 4 ha or more. Plates 1 and 2 give a general impression of the minimum size of water body included in the count. Loch Lomond (plate 3) is the lake with the largest area in Great Britain (excluding Ireland). Loch Brandy (plate 4) is one of Scotland's smaller lakes.

Sample counts on 30 by 30 km grid squares were also undertaken, the location of the sample areas being shown on Figure 1 and the results in Table 1. There is some indication that the original count misses more waters in lowland Britain than in the upland areas. The mean percentage counted in upland England and Wales is 12.8 and, in Scotland, 14.5. It is assumed that the latter two figures are not really different and

Plate 1 Scales Tarn near Keswick (not shown on 1:125000 map (1.25 ha)) (Hydrometric Area 75; Grid ref 3325 5283) Photograph R. H. Britton

Plate 2 Loch Fiart, Lismore Island. Surface area 0.13Km^2 (Hydrometric Area 89; Grid ref 1808 7377) Photograph P. S. Maitland

the mean value, viz 13.7, is taken to apply throughout the uplands. Since so few water bodies are included in the original count on lowland Britain (sample areas 7, 9, 10 and 11), the figures for percentage counted are very variable and it is better to estimate the uncounted waters using the ratio of the mean numbers in the samples, viz, 1.9 per cent.

The estimated numbers of waters that are not included in the count are listed in Table 2. The difference in the proportion of uncounted waters in lowland areas compared to the uplands is what might have been expected because of the large numbers of ponds, ornamental lakes etc, and the few large lakes and reservoirs in lowland areas.

Comparison between the two map scales in the Outer Hebrides (Hydrometric Area 106) and in some parts of the mainland in north west Scotland presents some difficulties. There are large numbers of small water bodies in this Area – 79 were counted in a single 10 by 10 km grid square and Area 106 contains one-fifth of the total waters in Britain. As a result, there are considerable cartographic differences between the two map scales and, in particular, it was observed that what was plotted as a number of separate, small lochans on 1:63360 maps often appeared as a single larger water body on the quarter inch maps.

Table 2 Estimated numbers of water bodies not included in the original count

	Numbers not included in original count	Numbers in original count	Total numbers
Scotland	27672	3788	31460
Upland England and Wales	6810	936	7746
Total upland	34482	4724	39206
Lowland England and Wales	41105	781	41886
Great Britain totals	75587	5505	81092

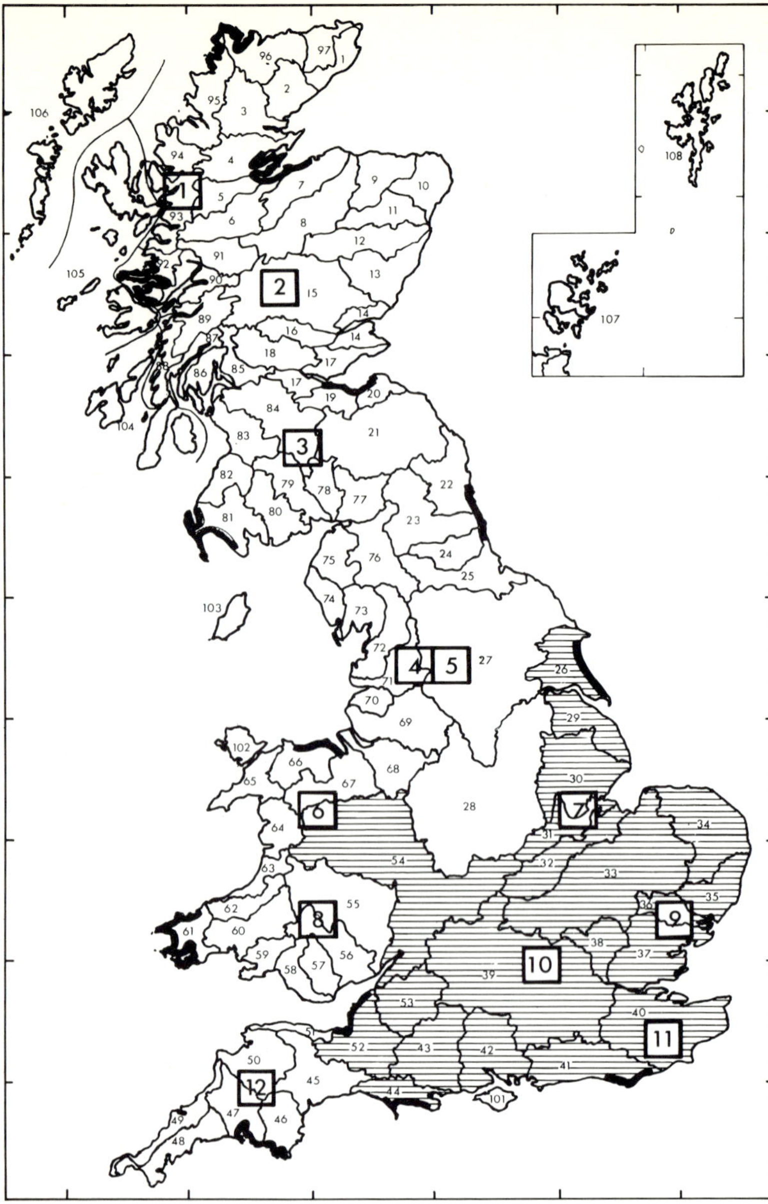

Figure 1 *The division of Great Britain into Hydrometric Areas.* The numbers refer to the Hydrometric Area number and the hatching indicates the extent of lowland Britain as used to calculate overall totals. For this purpose, lowland Britain is defined as those Hydrometric Areas whose solid geology consists largely of sedimentary rocks younger than the Permian. The squares are the sample areas used to check the adequacy of the lake and reservoir count and the marked lengths of coastline are the sample lengths used to check the river count

RESERVOIRS

An attempt has been made to distinguish between natural and artificial water bodies, except in the case of the class containing waters with an area less than .25 km². The definition of artificial has been limited here to water bodies formed by the construction of a dam. This means that, particularly in lowland areas, a number of waters such as clay and gravel pits have been counted as natural when they are, in fact, artificial. The reason for this restriction is purely practical – it is often impossible to decide from an examination of maps alone if the water is lying in a natural depression or not.

In England and Wales, the original data sheets were examined and all sites named as reservoirs were counted as artificial. Larger scale (1:63360) maps of all other sites were examined and these were counted as constructed reservoirs if a dam or artificial embankment was marked. A different procedure was used in Scotland because of the very much larger number of water bodies of this size and the existence of tabulated information. Reservoirs used for water supply are listed by the British Waterworks Association (1972). The list of waters affected by electric power developments was supplied by the North of Scotland Hydro-Electric Board (Pers. Comm.). Tables published by the Scottish Development Department (1973) indicate the reservoirs supplying canals in Scotland and other hydro-electric reservoirs operated by public authorities. In addition, there are a few privately owned hydro-electric reservoirs supplying aluminium works in the West Highlands. Further information on the numbers, distribution and type of reservoirs used for public water supply is given by Collinge and Davis (1975). They indicate a total of 521 water supply reservoirs, but this total includes some whose maximum surface area is less than 4 ha. In addition, 74 water bodies are affected by the operations of the North of Scotland Hydro-Electric Board, but not all of these are reservoirs created by a dam. These figures suggest that at least 10 per cent of the 5505 water bodies are reservoirs.

It must be emphasised that no absolute list of reservoirs can be made and that reservoir data is only included to give a more complete picture of water resources in Britain. The proportions of reservoirs in the different size classes, other than the smallest, are listed in Table 3.

Plate 3 Loch Lomond. Surface area 71.1Km² (Hydrometric Area 85; Grid ref 2350 6993) Photograph P. S. Maitland

Plate 4 Loch Brandy,' Glen Clova. Surface area 0.27Km² (Hydrometric Area 13; Grid ref 3339 7754) Photograph K. H. Morris

Table 3 The number of lakes and reservoirs in different size classes

		England	Wales	Scotland	GB total
<.25 km²	(a)	1224	219	2973	4416
.25–.50 km²	(a)	141	14	417	572
	(b)	57	43	8	21
.50–1.0 km²	(a)	57	10	209	276
	(b)	70	80	10	25
1–2 km²	(a)	17	9	83	109
	(b)	76	89	19	34
2–4 km²	(a)	9	5	54	68
	(b)	67	100	11	25
4–8 km²	(a)	7	3	27	37
	(b)	29	100	59	57
8–16 km²	(a)	2	—	12	14
	(b)	50	—	41	43
16–32 km²	(a)	—	—	9	9
	(b)	—	—	33	33
>32 km²	(a)	—	—	4	4
	(b)	—	—	25	25
Total		1457	260	3788	5505

(a) Total number of water bodies in the size class.
(b) Percentage of that number estimated to be reservoirs.

MEAN AND MAXIMUM DEPTHS

It was originally intended that the measured mean depth should be entered in the data files for all water bodies with a surface area greater than 2 km². For the majority of waters, this information can be obtained from the same sources as those used for water surface area. This was not always possible and also, in order to calculate the total volume of water held in the lakes and reservoirs of Britain, is necessary to devise methods of estimating the mean depth of any water body.

Gorham (1958) has shown that, for the Scottish lochs surveyed by Murray and Pullar (1910), there is a reasonable correlation between area and mean depth if the lochs are divided into rock and drift basins. Similar relations can be defined for reservoirs in any part of the country created by the construction of dams, if the topography is taken into account. The regression equations are listed in Table 4 and plotted on Figure 2.

Table 4 Mean depth: area regression

$\bar{D} = CA^m$ where $\bar{D}$ is the mean depth (m) and A is the area (km^2)

Type of water body	constant, C	exponent m	corr. coeff.	no of points
Gorham rock basin	8.30	.444	.68	262
Gorham drift basin	2.91	.153	.24	137
Reservoirs in defined valleys	13.55	.284	.67	28
Reservoirs in rolling topography	4.01	.234	.68	11

In the absence of similar data for England and Wales, it is necessary to make some assumptions. The first is that the Gorham regressions can be applied to rock and drift basins in upland England and Wales (as defined on Figure 1). Although such areas are topographically similar to Scotland, there are clearly errors, particularly in the Severn and Trent catchments (Areas 26 and 54) which are part upland and part lowland.

The other assumption is that the Gorham regression for drift basins can be applied to all natural waters in the lowlands. The admittedly doubtful justification for this assumption is contained on Figure 2. The steep face of a dam means that a reservoir will have greater mean depth for the same surface area than a natural lake. The area-mean depth curve for a lowland lake, therefore, must lie below that for a reservoir in rolling topography. In the absence of other data, the drift basin curve has been used as an approximate relationship. For the purpose of estimating volumes, the smallest size class (0.04–0.25 km^2) has not been separated into different types of water body and the mean depth appropriate to the geometric mean of the class, viz, 0.10 km^2, is taken to be 2.5 m (see Figure 2). For all water bodies greater than 2 km^2 the volumes of the individual lakes and reservoirs have been calculated, using, in a few cases, the mean depths estimated from the regression equations rather than field survey data. The total volumes appropriate to the three classes 0.25–0.50, 0.50–1.0 and 1–2 km^2 remain to be determined.

The problem is that, although the total number of water bodies and the number of reservoirs is known, the proportions of different types of water body is not known. The only evidence is

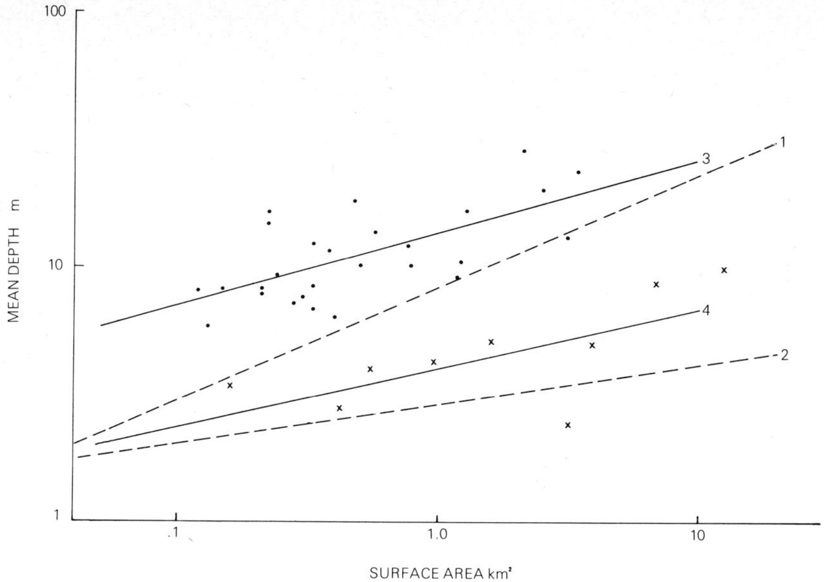

Figure 2 *The regression of mean depth on water surface area.* Lines 1 and 2 are the Gorham regressions for rock and drift basins respectively. Line 3 and the dots define the regression for reservoirs in defined valleys while line 4 and the crosses represent reservoirs in rolling topography

that, in the Bathymetrical Survey of Scottish lochs where Gorham (1958) identifies 262 rock basins and 137 drift basins, there was a ratio of 1.91 rock basins to one drift basin. Following the earlier argument, this ratio is assumed to apply to natural waters throughout upland Britain and to the ratio of reservoirs in defined valleys to those in rolling topography. Table 5 shows the weighted mean depths of the geometric means of the three classes on the basis of the Gorham ratio being applicable to upland areas and the relationships for drift basins and reservoirs in rolling topography being used in the lowlands.

These calculations are obviously approximate, but it is believed that they give a more accurate estimate of the volumes in the three classes than would result from combining all types of water without regard for their distribution throughout the country, and it is believed that the relative volumes in each class are reasonably correct.

The maximum depth was also recorded in the data files for all lakes and reservoirs with a surface area exceeding 2 km², and the procedure was similar to that for mean depth, ie surveyed depths were used wherever possible, but, where these were not available, estimated values were included. Gorham (1958) again provides the main source of information and he demonstrates the close correlation between mean and maximum depth for both rock and drift basins. A similar analysis was done for all types of reservoir. The results, shown in Table 6 and Figure 3, indicate that topography has little effect on the form of the relationships.

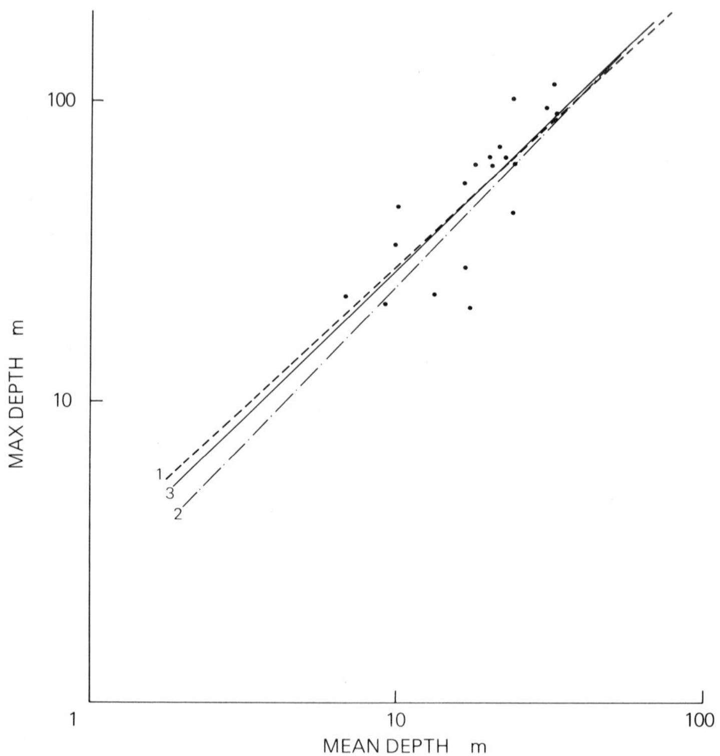

Figure 3 *The regression of maximum depth on mean depth.* Lines 1 and 2 are the Gorham regressions for rock and drift basins respectively. Line 3 and the dots define the regression for all types of reservoirs.

Table 5 Weighted mean depths (m) used to calculate total volumes

	size classes km²		
	.25–.50	.50–1.0	1.0–2.0
Uplands natural lakes	4.27	5.62	7.41
reservoirs	7.71	9.33	11.31
Lowlands natural lakes	2.48	2.76	3.07
reservoirs	3.15	3.70	4.35

Table 6 Maximum depth: mean depth regression

$D_{max} = K\bar{D}^n$ where D_{max} and $\bar{D}$ are the maximum and mean depths (m)

Type of water body	constant, K	exponent, n	corr. coeff.	no of points
Gorham rock basin	3.13	.955	.95	262
Gorham drift basin	2.14	1.06	.92	137
Reservoirs (all types)	2.83	.992	.88	22

2 The river survey

The lake and reservoir count was based on the 1:250000 (four miles to the inch) maps, but much of the river detail on these maps is obscured by additional man-made features. It was, therefore, decided to base the count on the 1:625000 (ten miles to the inch) physical maps on which these additional features are omitted. Preliminary investigations had already shown the lack of consistency between maps of different scale in delineating river networks. Since the task of examining the entire country using detailed maps, eg, at a scale of 1:50000, is enormous, the change of map scale was considered justified.

The only measure of size used in the river survey is stream order as defined by Strahler (1957). A stream without tributaries is defined as first order, the stream below the confluence of two first orders is a second order and so on (see Figure 4). The figure shows that the order number of the outfall stream is very dependent on the chance inclusion or omission of a first order tributary. The omission from the map of stream A on the figure would reduce the outfall order number from 4 to 3. The uncertainty in stream ordering can be seen when counts on different map scales are compared (see below). For an indication of actual sizes of streams and rivers corresponding to the various streams order numbers on 1:625000 maps see

Plates 5–14 inclusive. Plates 15 and 16 illustrate the size of stream omitted in the original river count.

It is a characteristic of river networks that there is a linear relationship between the stream order number and the logarithm of the number of streams in each order class (see, for example, Leopold, Wolman and Miller, 1964). The slope of the line (order number being on the X-axis) is referred to as the bifurcation ratio.

THE ORIGINAL COUNT

For each river system entering the sea, the following information was tabulated: the Hydrometric Area; the National Grid Reference of the mouth; the name (if marked); the order number of the outfall stream; the total number of streams of various orders associated with each sea inflow. The total numbers of river systems and associated streams are listed for each Hydrometric Area in Appendix 2. The total number of river systems is 1445 with which are associated 7835 streams.

MAP SCALE TRANSFORMATIONS

For 28 river systems distributed over the whole country, stream order analysis was repeated on the 1:63360 (one inch to one mile) maps. In general, the change of map scale increases the order number of the outfall stream by a factor of 2 but not consistently (see Figure 5).

If the order number of the sea inflow is known, the total number of streams in each lower order can be estimated if the bifurcation ratio is known. Figure 6 shows that the bifurcation ratio is not quite the same when estimated from maps of different scale. Only systems occurring as fifth order or above on the 1:63360 maps were included in the regression in order to obtain meaningful estimates of bifurcation ratio from the 1:625000 maps. The average value was 3.75 on the 1:63360 maps and 3.32 on the smaller scale.

This examination of the effects of map scale makes it possible to estimate what the information obtained from the 1:625000 maps means in terms of an equivalent examination of the river systems on 1:63360 maps. Because of the statistical nature of the transformations (only two thirds of the difference is directly explained by map scale changes), such a comparison becomes less and less accurate as the area or length of coast being considered is reduced.

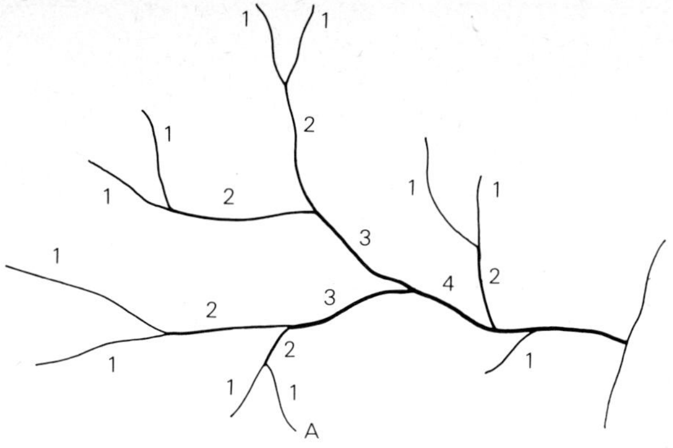

Figure 4 *The definition of stream order number*

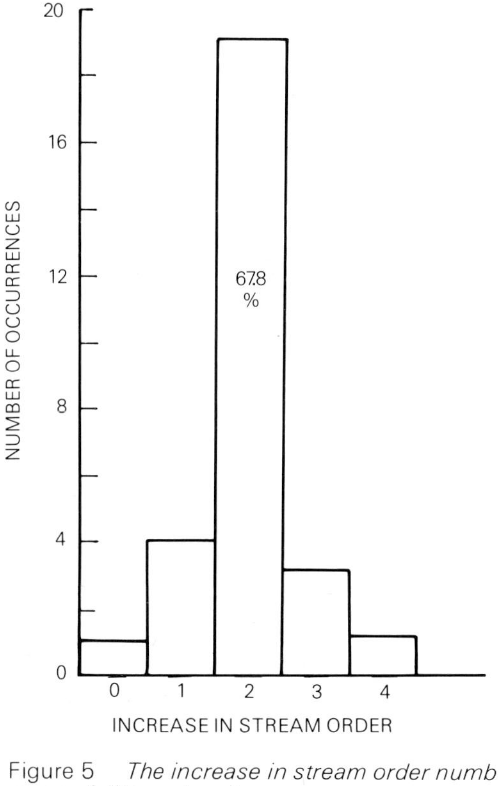

Figure 5 *The increase in stream order number when comparing maps of different scale*

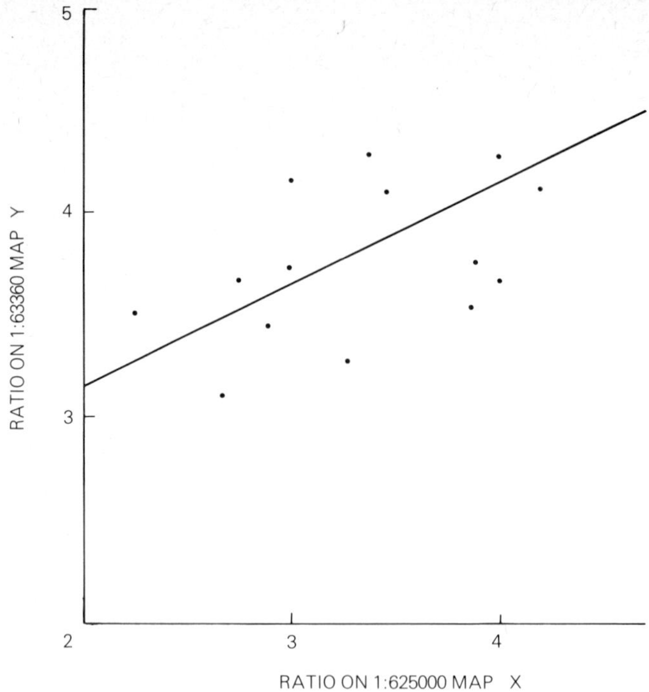

Figure 6 *The influence of map scale on bifurcation ratio.* Y=.503X + 2.14 (n=14, r=.822)

Plate 5 River Dove, near Hartington. 1st order (Hydrometric Area 28; Grid ref 4128 3587) Photograph R. H. Britton

Plate 6 Water of Tarf, tributary of the North Esk. 1st order (Hydrometric Area 13; Grid ref 3487 7827) Photograph A. A. Lyle

Plate 7 River Teifi, near Tregaron. 2nd order (Hydrometric Area 62; Grid ref 2680 2625) Photograph R. H. Britton

Plate 8 River Tilt, Glen Tilt. 2nd order (Hydrometric Area 15; Grid ref 2953 7758) Photograph K. H. Morris

Plate 9 River Ericht near Blairgowrie. 3rd order (Hydrometric Area 15; Grid ref 3220 7442) Photograph K. East

Plate 10 River North Esk at Edzell. 3rd order (Hydrometric Area 13; Grid ref 3602 7690) Photograph A. A. Lyle

Plate 11 River Tay at Dunkeld. 4th order (Hydrometric Area 15; Grid ref 3012 7425) Photograph D. H. Jones

Plate 12 River Wye below Hereford. 4th order (Hydrometric Area 55; Grid ref 3584 2272) Photograph R. H. Britton

Plate 13 River Tay 12 Km upstream of Perth. 5th order (Hydrometric Area 15; Grid ref 3149 7370) Photograph K. H. Morris

Plate 14 River Tay at Perth. 5th order (Hydrometric Area 15; Grid ref 3121 7242) Photography K. H. Morris

RIVER SYSTEMS OMITTED FROM THE ORIGINAL COUNT

In order to estimate the likely total number of river systems on 1:63360 maps, it is necessary to know what rivers are omitted from the original count. Thirteen lengths of coastline were examined (see Figure 1) and the number of additional, previously uncounted, river systems noted. Figure 7 shows that it is possible to estimate the number of additional systems when the number in the original count is known.

These sample lengths of coastline accounted for 839 river systems not included in the original count. The order number of the outfalls of these systems are shown in Table 7. There are approximately six times as many systems on the 1:63360 maps compared to the original count, but 93 per cent of these are of first or second order which is consistent with the previous stream order transformation.

3 The extent and distribution of freshwaters in Great Britain

The national totals of the numbers of standing waters in the different size classes are shown on Table 3, as is the percentage of those numbers estimated to be reservoirs. The most obvious features are the very large numbers in Scotland (over a fifth of the British total are in the Outer Hebrides alone) and the high proportion of the larger waters in England and Wales that are estimated to be reservoirs.

The dominance of Scotland in terms of numbers is much less pronounced when the uncounted waters estimated to be marked on the 1:63360 maps are taken into account. Referring to Table 2, the total for England and Wales is 49632 compared to 31460 for Scotland.

The national totals for standing waters are exact since it is possible to assign each water to a particular country. This is not the case with river networks which cross boundaries and, in some cases, where the centre line of the river is the national boundary. The national totals for river systems and streams in Table 8 are, therefore, based on complete Hydrometric Areas. This allocation can lead to some error. For example, in the Severn catchment (Hydrometric Area 54), there are 64 streams above the confluence of the Severn and the Vyrnwy which are included in the total for England, but which are, in fact, in Wales.

A feature of Tables 3 and 8 is that the proportions of the overall totals for both lakes and river systems in the different countries are virtually the same. England has 26 per cent of the counted lakes and river systems, the figures for Wales are 5 per cent for lakes and 8 for rivers while in Scotland the proportions are 69 and 66 respectively.

By combining the average effect of map scale transformation with the data for uncounted systems, it is possible to estimate the total number of river systems in Great Britain (see Table 9). The total number of systems in the original count is 1445 so that the number of uncounted systems, using the regression of Figure 7, is 8638, giving an overall total of 10083. The uncounted total is assumed to be distributed between different orders according to the percentages of Table 7. The form of the regression in Figure 7 is such that its direct application to the Great Britain total is permissible.

The mean bifurcation ratio for all rivers of third order or above that were examined on the 1:63360 maps is 3.86, with a standard deviation of 0.96 (this figure differs slightly from that obtained from the restricted sample used to establish Figure 6). Using this figure, the approximate number of streams associated with these 10083 river systems can be calculated. The total number is 194674 and their distribution between stream order sizes and between systems of different order is shown on Table 10.

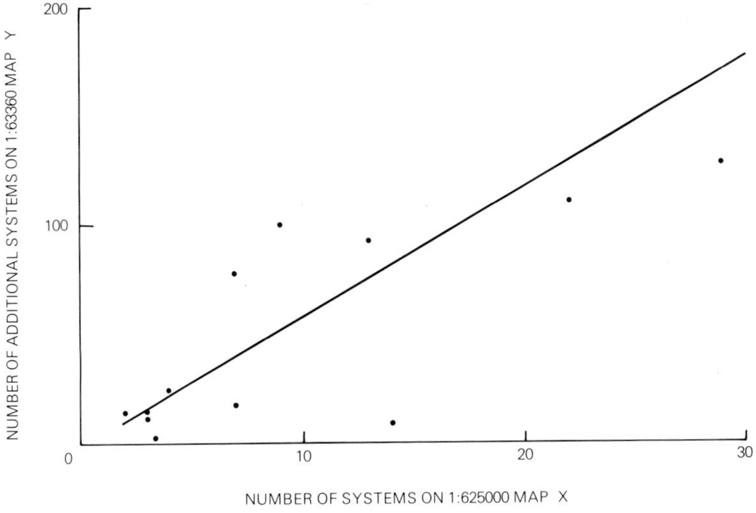

Figure 7 *The number of additional river systems on 1:63360 maps in terms of the number on 1:625000 maps (sample lengths of coast).* Y=5.98X − 2.95 (n=13, r=.833)

Table 7 Stream order (1:63360 scale) of systems omitted from original count (sample lengths of coast)

	1st	2nd	3rd	4th	Total
Number	566	214	58	1	839
Per cent of total	67.5	25.5	6.9	1	100

Table 8 River systems and streams in original count

	Order number					
	1	2	3	4	5	Total
(a) River systems						
England[1]	194	106	44	26	1	371
Wales[2]	65	37	16	6	–	124
Scotland[3]	563	296	69	19	3	950
Great Britain total	822	439	129	51	4	1445
(b) Streams						
England[1]	2328	615	147	28	1	3119
Wales[2]	577	143	33	6	–	759
Scotland[3]	3061	733	134	26	3	3957
Great Britain total	5966	1491	314	60	4	7835

[1]Hydrometric Areas 22–54, 68–76, 101, 103
[2]Hydrometric Areas 55–67, 102
[3]Hydrometric Areas 1–21, 77–97, 104–108

The geographical distribution of counted freshwaters is displayed in more detail in a series of maps (Figure 8). Figures 8a and 8b are derived by dividing the number of standing waters and streams in a Hydrometric Area by its total area. Figure 8c displays the area of lakes and reservoirs divided by the number of square kilometres in the Area while Figure 8d represents the area of standing water divided by the number of lakes and reservoirs.

As well as emphasising the obvious dominance of the Highlands and Islands of Scotland, the maps show that there are areas which are always in the lowest class for both standing and running waters. These areas are in the North East of Scotland (Areas 9 and 11), East Anglia (Areas 35 and 36) and Wessex (Areas 43 and 53).

The official estimates of all inland waters in Britain are given in Table 11 (Central Office of Information, 1977). The 2404 km² of inland water represents 1.04 per cent of the total area of Britain.

An obvious feature of Table 11 is the difference in the proportions of the total areas represented by the lake and reservoir count in England and Wales compared to Scotland. The sample comparisons of different map scales showed that the 1:250000 maps omitted a very large number of small water bodies in lowland Britain, while a large fraction of the water area in Scotland is made up of larger lochs which are included in the count. Despite these and other physical explanations – existence of the extensive drainage network in lowland England – it is believed that this difference is also due to the cartographic errors in the Outer Hebrides, and perhaps elsewhere, already referred to.

Table 9 Estimated total number of river systems in Great Britain

	Stream order on 1:63360 map						
	1	2	3	4	5	6	7
Transformed total from original count	—	—	822	439	129	51	4
Estimated number of uncounted systems	5830	2202	596	10	—	—	–
Total	5830	2202	1418	449	129	51	4

If it is assumed that the standing water count includes all lakes and reservoirs with a surface area greater than 4 ha, then the less than .25 km^2 class includes at least two classes on the logarithmic scale ie .0625–.125–.25 km^2. If allowance is made for this and the uncertainty about the numbers in the Outer Isles, the histograms of Figure 9 suggest that there is some pattern in the quantities and extent of freshwaters in Britain.

There is a more or less constant decrease in the logarithms of the numbers of standing waters with decreasing size class, the water surface area represented by each class is roughly constant while the distribution of volumes is dominated by the few, very large Scottish lochs. The total volume of all standing waters counted is estimated to be $38.3 \times 10^9 m^3$, the total for England and Wales is about one tenth of this, less than the volume of Loch Ness alone.

The linear decrease in the logarithms of the numbers with increasing size (stream order number) appears to apply to river systems as well as standing waters except that the number of rivers of the largest size is somewhat less than might have been expected. The number of streams associated with these systems is distributed rather erratically.

Cartographic inconsistencies mean that the use of stream order as the basis for defining river size is not completely satisfactory and that it is unwise to use this method to come to any detailed conclusions about the relative size of particular rivers. True river size is best defined by the flow rate, making use of the relationships between flow and river dimensions (see, for example, Leopold, Wolman and Miller, 1964).

Table 10 The approximate number of streams in Great Britain

River system order number	Stream order number 1	2	3	4	5	6	7	Total numbers in each system group
1	5830							5830
2	8500	2202						10702
3	21128	5473	1418					28019
4	25823	6690	1733	449				34695
5	28638	7419	1922	498	129			38606
6	43703	11322	2933	760	197	51		58966
7	13231	3428	888	230	60	15	4	17856
Totals	146853	36534	8894	1937	386	66	4	194674

Table 11 Water surface areas in Great Britain (km^2)

	Area including inland water	Area of inland water	Water areas as percentage of land area	Area included in lake and reservoir count	Per cent of inland water included in lake count
England	130375	675	.52	322.5	47.8
Wales	20766	125	.60	73.9	59.0
Scotland	78774	1604	2.04	1527.9	95.3
Great Britain	229915	2404	1.04	1924.4	80.0

Plate 15 Allt a Ghlinne Bhig. Glen Tilt (not shown on 1:625000 map) (Hydrometric Area 15; Grid ref 3035 7800)
Photograph A. A. Lyle

Plate 16 and cover Gairney Water at Powmill (not shown on 1:625000 map) (Hydrometric Area 17; Grid ref 3012 6981)
Photograph K. East

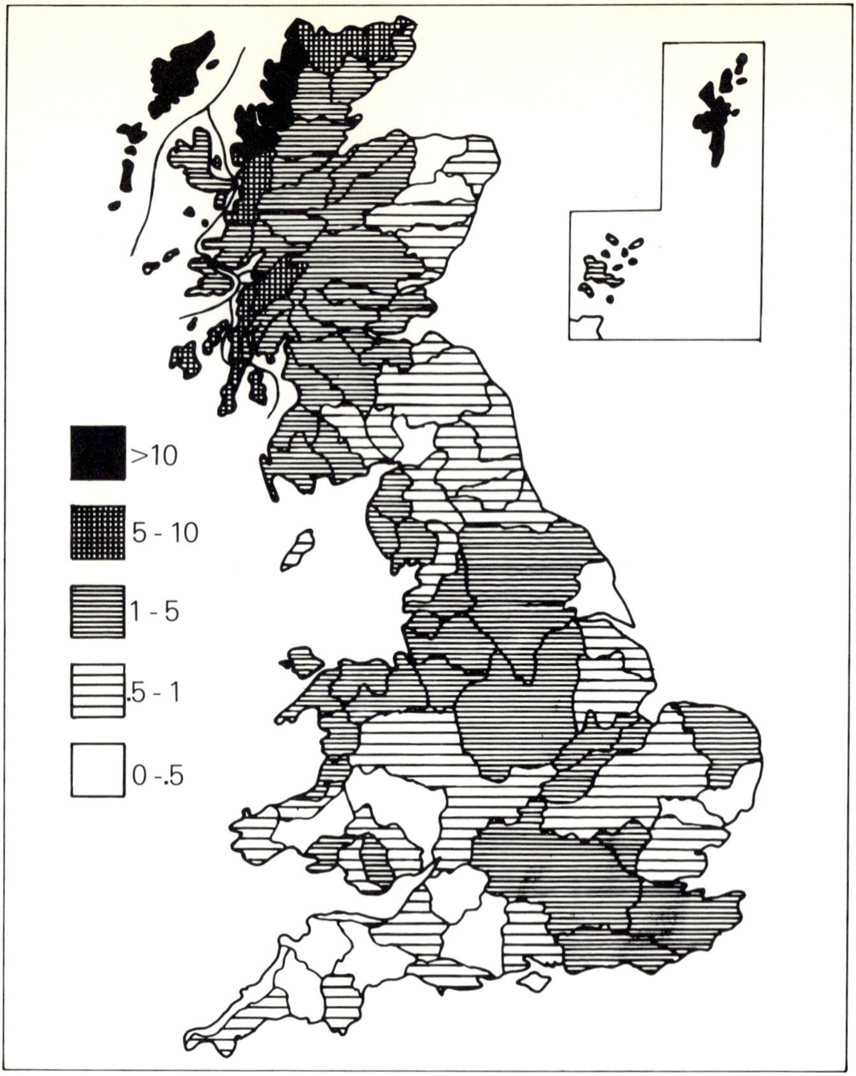

Figure 8 *The extent and distribution of freshwaters by Hydrometric Area.* (a) The number of standing waters per 100 km^2;

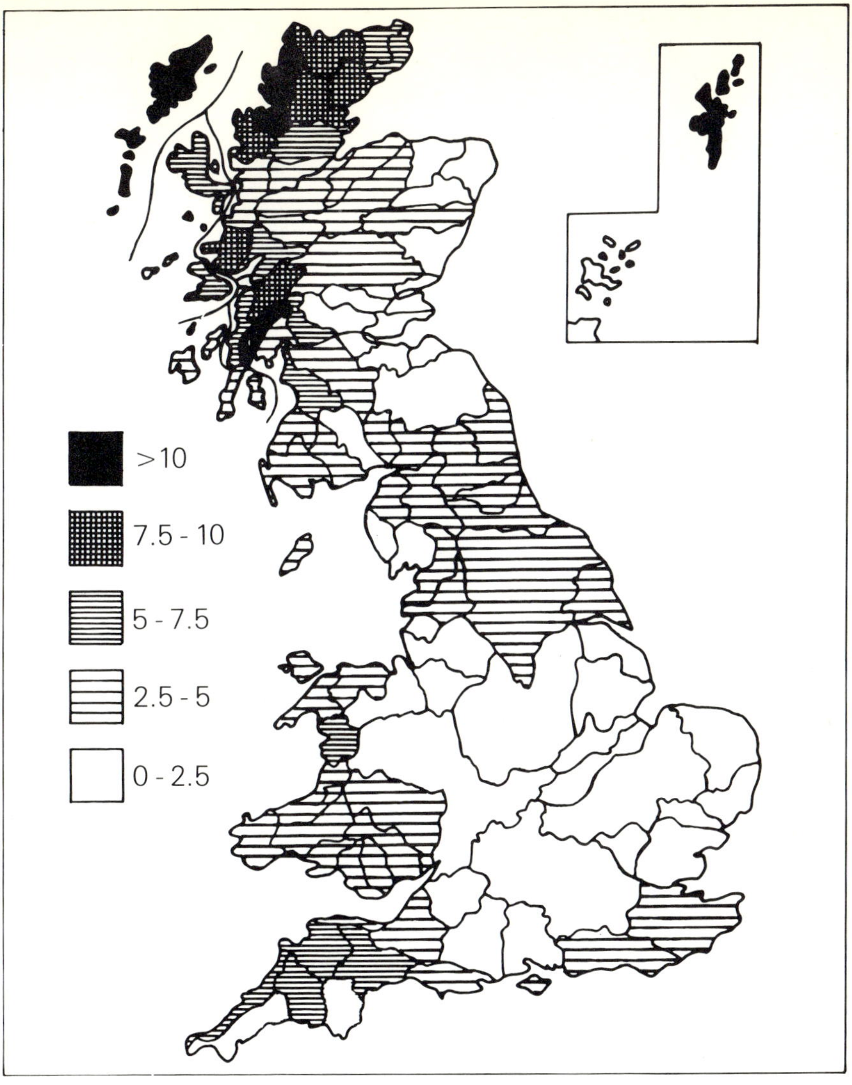

Figure 8 (b) The number of streams per 100 km^2;

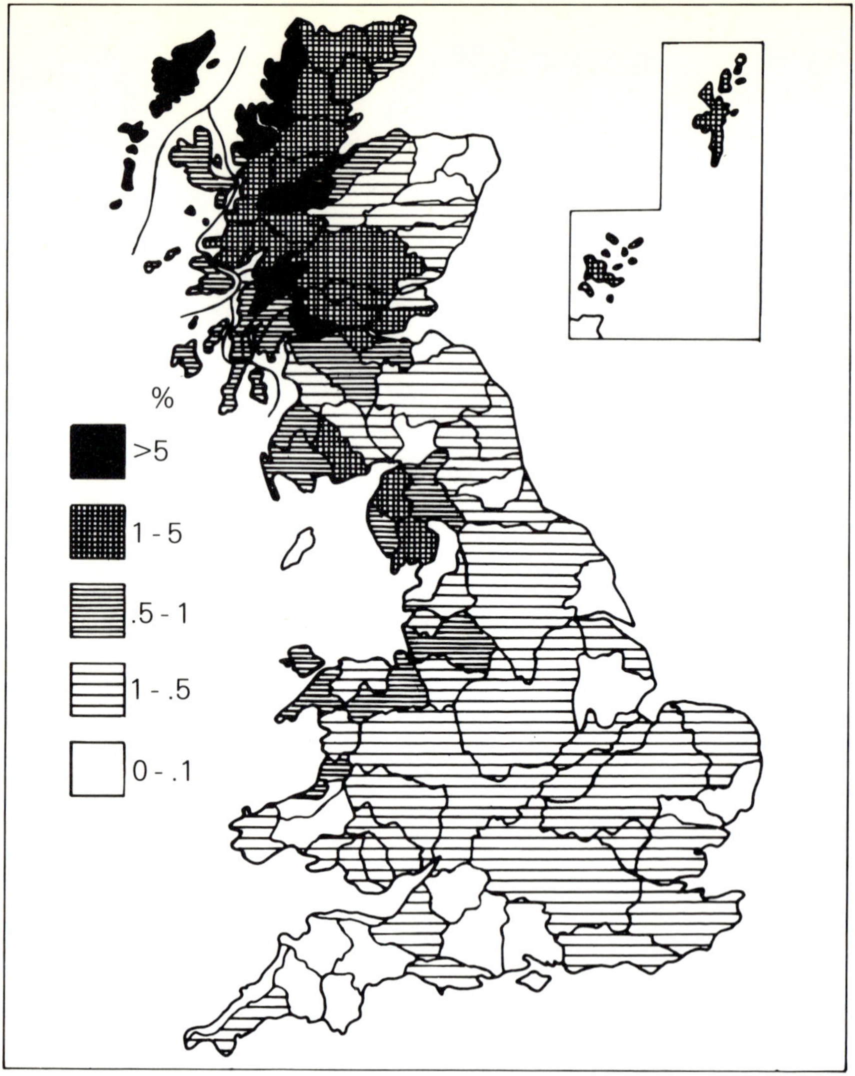

Figure 8 (c) The percentage area of standing water;

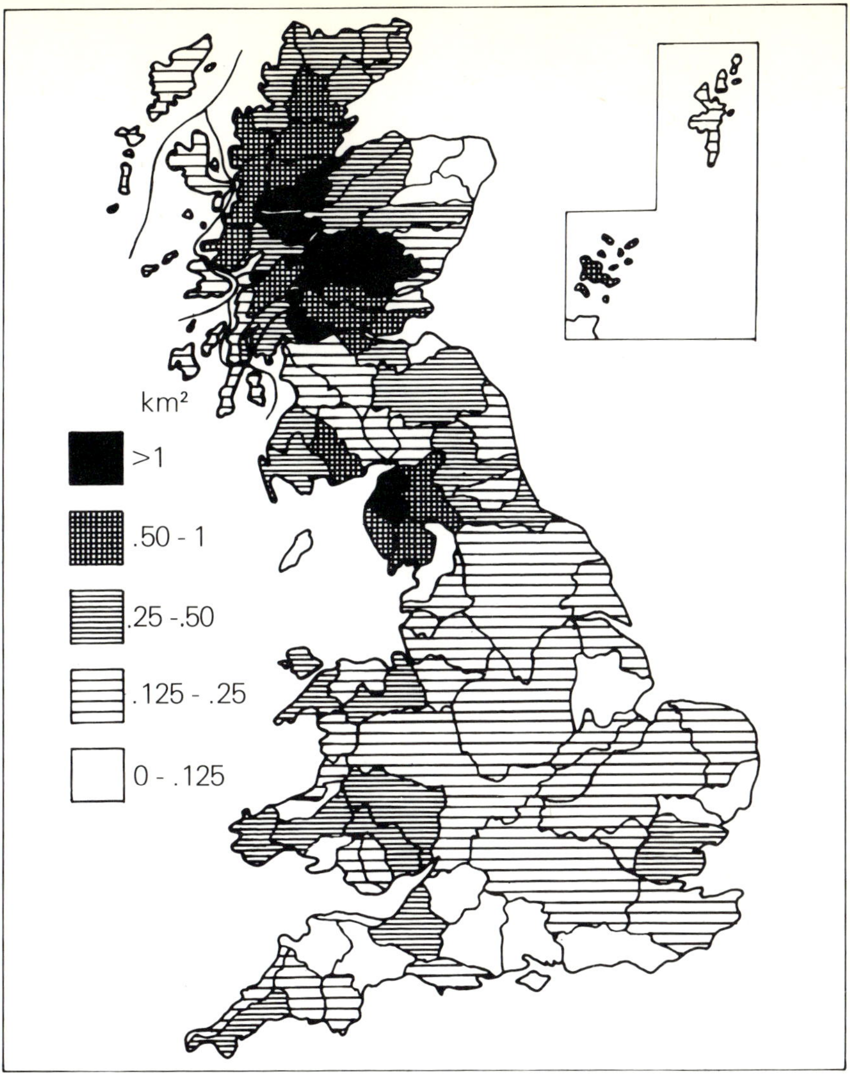

Figure 8 (d) The mean area of standing waters.

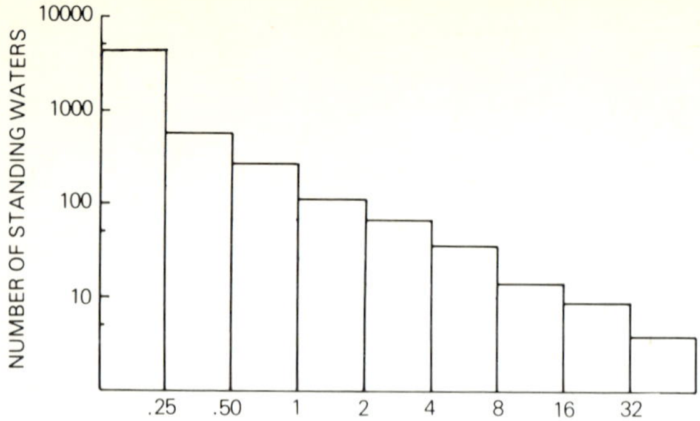
NUMBER OF STANDING WATERS
10000
1000
100
10
.25
.50
1
2
4
8
16
32

(a)

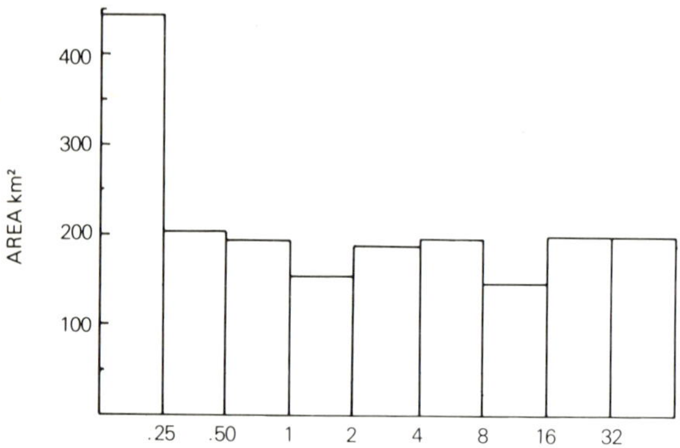
AREA km²
400
300
200
100
.25
.50
1
2
4
8
16
32

(b)

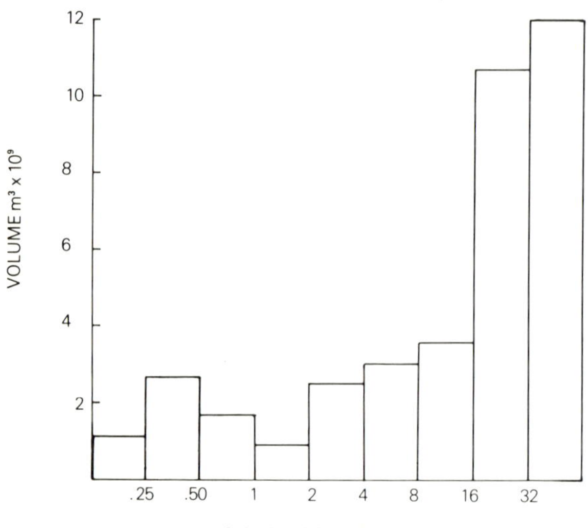
VOLUME m³ x 10⁹
12
10
8
6
4
2
.25
.50
1
2
4
8
16
32
SIZE CLASS INTERVAL km²

(c)

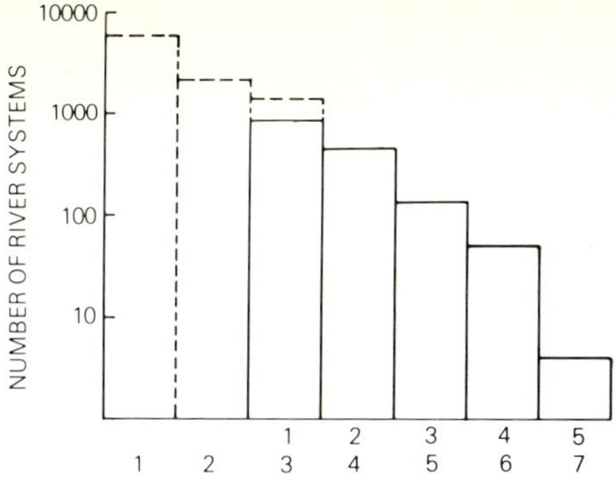

(d)

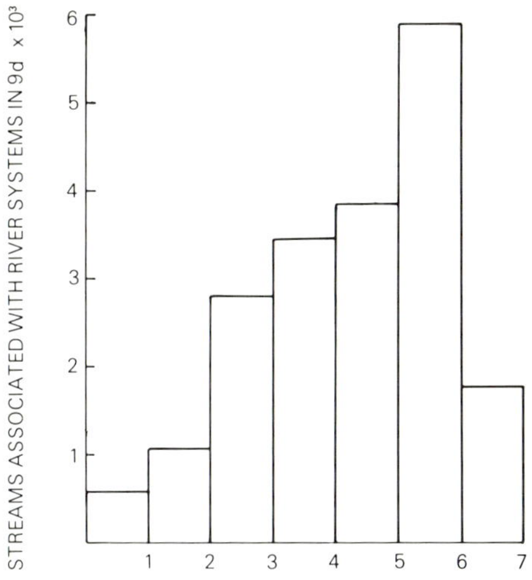

(e)

Figure 9 *The extent and distribution of freshwaters by size classes.* (a) The number of standing waters in the different size classes; (b) The water surface area in the different size classes; (c) The water volumes in the different size classes; (d) The number of river systems of different order number. (The full lines and the upper stream order numbers refer to the original count. The broken lines and the lower stream order numbers refer to the estimated numbers on 1:63360 maps); (e) The number of streams associated with the river systems in (d).

Appendix 1
The numbers of lakes and reservoirs in each Hydrometric Area subdivided according to water surface area

Hydro Area	Numbers of lakes and reservoirs: area classes in km²								
	<.25	.25–.50	.50–1.0	1.0–2.0	2.0–4.0	4–8	8–16	16–32	>32	Total
1	12	3	3	–	1	–	–	–	–	19
2	33	11	4	1	5	–	–	–	–	54
3	55	11	4	6	–	–	–	–	1	77
4	59	11	6	5	2	4	1	–	–	88
5	35	4	4	1	1	1	2	–	–	48
6	63	7	5	5	–	3	1	1	1	86
7	21	6	1	–	1	1	–	–	–	30
8	21	5	8	3	–	–	–	–	–	37
9	–	–	–	–	–	–	–	–	–	0
10	7	1	–	–	1	–	–	–	–	9
11	–	–	–	–	–	–	–	–	–	0
12	14	2	2	1	1	–	–	–	–	20
13	12	1	2	–	–	–	–	–	–	15
14	7	2	1	–	–	–	–	–	–	10
15	59	12	12	6	3	3	–	3	–	98
16	10	2	–	2	–	–	1	–	–	15
17	35	8	4	–	1	–	1	–	–	49
18	24	2	6	2	4	1	1	–	–	40
19	15	1	2	3	–	–	–	–	–	21
20	5	–	–	–	–	–	–	–	–	5
21	29	6	–	2	1	–	–	–	–	38
22	17	1	1	–	–	–	–	–	–	19
23	13	3	3	–	–	1	–	–	–	20
24	6	1	–	–	–	–	–	–	–	7
25	13	1	3	1	–	–	–	–	–	18
26	7	1	1	–	–	–	–	–	–	9
27	119	29	10	–	–	–	–	–	–	158
28	108	11	9	4	1	–	–	–	–	133
29	15	1	1	–	–	–	–	–	–	17
30	25	–	–	–	–	–	–	–	–	25
31	15	1	–	1	–	–	–	–	–	17
32	34	7	1	–	1	–	–	–	–	43
33	60	4	1	–	–	1	–	–	–	66
34	45	6	1	–	–	–	–	–	–	52

35	4	–	–	–	–	–	–	–	–	4
36	2	–	–	–	–	–	–	–	–	2
37	28	–	–	–	1	1	–	–	–	30
38	25	5	2	1	–	–	–	–	–	33
39	168	18	6	3	1	–	–	–	–	196
40	62	6	4	1	–	–	–	–	–	73
41	37	2	–	–	–	–	–	–	–	39
42	21	–	–	–	–	–	–	–	–	21
43	11	1	–	–	–	–	–	–	–	12
44	10	1	–	–	–	–	–	–	–	11
45	4	–	–	–	–	–	–	–	–	4
46	8	2	–	–	–	–	–	–	–	10
47	5	–	1	–	–	–	–	–	–	6
48	12	–	2	1	–	–	–	–	–	15
49	3	1	–	–	–	–	–	–	–	4
50	6	–	–	–	–	–	–	–	–	6
51	3	–	–	–	–	–	–	–	–	3
52	13	–	3	1	–	–	–	–	–	17
53	10	–	–	–	–	1	–	–	–	11
54	98	9	1	1	1	1	–	–	–	111
55	14	–	2	2	1	–	–	–	–	19
56	11	1	–	3	–	–	–	–	–	15
57	13	–	1	1	–	–	–	–	–	15
58	6	–	1	–	–	–	–	–	–	7
59	10	–	–	–	–	–	–	–	–	10
60	6	–	–	1	–	–	–	–	–	7
61	7	–	–	1	–	–	–	–	–	8
62	8	–	–	–	–	–	–	–	–	8
63	22	1	–	–	1	–	–	–	–	24
64	18	–	1	–	–	–	–	–	–	19
65	34	6	3	–	–	1	–	–	–	44
66	22	4	1	–	–	–	–	–	–	27
67	23	–	–	1	1	1	–	–	–	26
68	35	4	–	–	–	–	–	–	–	39
69	71	19	3	–	–	–	–	–	–	93
70	9	2	2	–	–	–	–	–	–	13
71	35	–	–	1	–	–	–	–	–	36
72	13	–	–	–	–	–	–	–	–	13
73	26	1	2	1	–	1	1	–	–	32
74	9	2	–	–	2	–	–	–	–	13
75	8	1	–	1	2	2	–	–	–	14

Appendix 1 (cont'd)

Hydro Area	Numbers of lakes and reservoirs: area classes in km²								
	<.25	.25–.50	.50–1.0	1.0–2.0	2.0–4.0	4–8	8–16	16–32	>32	Total
76	14	1	–	–	1	–	1	–	–	17
77	1	1	–	–	–	–	–	–	–	2
78	7	1	1	–	–	–	–	–	–	9
79	10	3	1	–	–	–	–	–	–	14
80	22	11	4	1	–	2	–	–	–	40
81	27	7	8	–	–	–	–	–	–	42
82	19	4	4	–	–	1	–	–	–	28
83	22	3	1	–	–	–	–	–	–	26
84	63	14	8	2	1	–	–	–	–	88
85	16	1	–	1	1	–	–	–	1	20
86	21	3	2	1	–	1	–	–	–	28
87	15	2	–	1	1	–	–	–	–	19
88	70	2	1	2	–	–	–	–	–	75
89	43	10	–	–	3	–	–	–	1	57
90	34	3	–	1	–	–	1	–	–	39
91	20	1	5	1	2	2	–	2	–	33
92	44	7	2	2	–	–	–	1	–	56
93	73	8	3	2	–	–	–	1	–	87
94	81	13	7	6	2	–	–	1	–	110
95	313	49	18	7	8	1	2	–	–	398
96	163	20	6	2	2	3	–	–	–	196
97	36	11	5	1	1	–	–	–	–	54
101	1	–	–	–	–	–	–	–	–	1
102	14	2	1	–	1	–	–	–	–	18
103	4	–	–	–	–	–	–	–	–	4
104	119	3	4	1	1	–	–	–	–	128
105	125	13	4	2	1	1	–	–	–	146
106	910	104	53	8	8	2	1	–	–	1094
107	31	6	4	3	2	1	1	–	–	48
108	167	22	4	2	–	–	–	–	–	195
All Areas	4416	572	276	109	68	37	14	9	4	5505

Appendix 2
The total number of river systems and streams counted on 1:625000 maps

Hydro Area	No of systems						No of streams					
	Stream order number											
	1	2	3	4	5	Total	1	2	3	4	5	Total
1	6	1	3	1	–	11	48	12	5	1	–	66
2	6	–	3	1	–	10	89	22	5	1	–	117
3	3	7	2	–	1	13	141	40	8	2	1	192
4	4	2	2	1	–	9	84	15	4	1	–	104
5	2	–	–	1	–	3	39	7	3	1	–	50
6	–	–	–	1	–	1	60	14	3	1	–	78
7	1	2	1	1	–	5	46	13	3	1	–	63
8	–	–	–	1	–	1	72	16	5	1	–	94
9	6	1	1	–	–	8	25	5	1	–	–	31
10	7	2	1	–	–	10	27	5	1	–	–	33
11	–	–	1	–	–	1	22	4	1	–	–	27
12	–	–	–	1	–	1	42	11	2	1	–	56
13	7	1	1	1	–	10	38	8	3	1	–	50
14	6	3	–	–	–	9	12	3	–	–	–	15
15	–	–	–	–	1	1	124	26	6	2	1	159
16	–	–	1	–	–	1	18	5	1	–	–	24
17	6	2	1	–	–	9	15	4	1	–	–	20
18	3	–	–	1	–	4	36	10	4	1	–	51
19	1	–	2	–	–	3	12	4	2	–	–	18
20	3	1	–	–	–	4	8	1	–	–	–	9
21	1	2	1	–	1	5	89	27	8	3	1	128
22	4	4	3	–	–	11	42	10	3	–	–	55
23	–	–	–	1	–	1	67	19	3	1	–	90
24	1	–	1	–	–	2	25	7	1	–	–	33
25	5	4	–	1	–	10	66	16	2	1	–	85
26	5	4	3	–	–	12	39	12	3	–	–	54
27	5	1	1	3	1	11	247	62	19	5	1	334
28	2	–	–	1	–	3	81	23	8	1	–	113
29	8	6	–	–	–	14	25	6	–	–	–	31
30	1	–	1	–	–	2	20	4	1	–	–	25
31	–	–	–	1	–	1	15	4	2	1	–	22
32	–	–	1	–	–	1	19	5	1	–	–	25
33	1	1	–	1	–	3	55	17	5	1	–	78
34	3	1	3	–	–	7	38	8	3	–	–	49
35	10	3	–	–	–	13	17	3	–	–	–	20
36	1	–	1	–	–	2	8	2	1	–	–	11
37	8	7	2	–	–	17	48	14	2	–	–	64
38	–	–	1	–	–	1	18	5	1	–	–	24
39	–	–	1	1	–	2	133	37	9	1	–	180
40	4	2	2	2	–	10	114	33	9	2	–	158
41	4	3	2	3	–	12	102	32	8	3	–	145
42	13	7	1	–	–	21	30	10	1	–	–	41
43	1	–	1	1	–	3	55	14	4	1	–	74
44	3	5	2	–	–	10	32	9	2	–	–	43
45	3	3	2	1	–	9	90	19	5	1	–	115
46	4	5	2	–	–	11	57	22	2	–	–	71
47	4	6	1	1	–	12	71	20	4	1	–	96

Hydro Area	No of systems						No of streams					
	Stream order number											
	1	2	3	4	5	Total	1	2	3	4	5	Total
48	33	7	2	–	–	42	65	11	2	–	–	78
49	14	11	1	–	–	26	57	16	1	–	–	74
50	4	2	–	2	–	8	85	21	5	2	–	113
51	6	3	1	–	–	10	21	6	1	–	–	28
52	9	2	2	1	–	14	99	28	6	1	–	134
53	1	–	1	–	–	2	26	6	1	–	–	33
54	8	2	1	1	–	12	147	39	10	1	–	197
55	1	–	–	1	–	2	96	21	5	1	–	123
56	3	1	–	1	–	5	42	9	2	1	–	54
57	–	2	1	–	–	3	21	5	1	–	–	27
58	2	4	2	–	–	8	35	9	2	–	–	46
59	6	2	2	–	–	10	26	6	2	–	–	34
60	4	3	1	1	–	9	66	20	6	1	–	93
61	22	3	2	–	–	27	46	7	2	–	–	55
62	–	–	3	–	–	3	35	6	1	–	–	42
63	8	2	1	–	–	11	24	5	1	–	–	30
64	4	4	2	1	–	11	51	15	4	1	–	71
65	4	10	1	–	–	15	39	12	1	–	–	52
66	3	–	1	1	–	5	39	10	3	1	–	53
67	1	–	–	1	–	2	36	12	3	1	–	52
68	2	2	–	1	–	5	30	8	2	1	–	41
69	1	2	–	1	–	4	42	13	4	1	–	60
70	–	2	–	–	–	2	6	2	–	–	–	8
71	–	–	–	1	–	1	34	12	4	1	–	51
72	3	–	1	1	–	5	48	13	4	1	–	66
73	4	–	2	–	–	6	20	6	2	–	–	28
74	4	6	–	–	–	10	21	6	–	–	–	27
75	3	1	1	–	–	5	27	7	1	–	–	35
76	–	–	–	1	–	1	57	12	4	1	–	74
77	1	1	1	–	–	3	36	9	1	–	–	46
78	–	–	1	–	–	1	20	6	1	–	–	27
79	–	2	1	–	–	3	28	8	1	–	–	37
80	5	1	1	1	–	8	42	11	3	1	–	57
81	11	4	2	–	–	17	43	9	2	–	–	54
82	–	1	3	–	–	4	28	8	3	–	–	39
83	1	5	2	1	–	9	47	13	4	1	–	65
84	–	1	2	1	–	4	66	18	4	1	–	89
85	2	–	–	1	–	3	40	10	3	1	–	54
86	20	5	2	–	–	27	43	9	2	–	–	54
87	7	12	2	–	–	21	56	16	2	–	–	74
88	4	11	2	–	–	17	57	17	2	–	–	76
89	7	3	–	1	–	11	58	12	2	1	–	73
90	14	11	1	–	–	26	47	13	1	–	–	61
91	–	–	–	1	–	1	43	9	2	1	–	55
92	19	11	2	–	–	32	71	17	2	–	–	90
93	24	9	2	–	–	35	64	13	2	–	–	79
94	12	13	2	–	–	27	64	20	2	–	–	86
95	28	17	9	–	–	54	183	43	9	–	–	235

Hydro Area	No of systems						No of streams					
	Stream order number											
	1	2	3	4	5	Total	1	2	3	4	5	Total
96	14	11	3	2	–	30	132	35	7	2	–	176
97	7	2	–	1	–	10	42	9	2	1	–	54
101	8	1	–	–	–	9	11	1	–	–	–	12
102	7	6	–	–	–	13	21	6	–	–	–	27
103	4	3	1	–	–	8	18	5	1	–	–	24
104	37	42	1	–	–	80	153	44	1	–	–	198
105	84	29	1	–	–	114	163	31	1	–	–	195
106	92	35	5	–	–	132	236	46	5	–	–	287
107	10	1	–	–	–	11	12	1	–	–	–	13
108	102	45	4	–	–	151	240	54	4	–	–	298

REFERENCES

British Waterworks Association. (1972). Amenity Use of Reservoirs in Scotland, Analysis of Returns, 1971, duplicated report.

Central Office of Information. (1974). Britain 1974. An official handbook. HMSO, London.

Collinge, V. K. and Davis, J. M. (1975). A survey of British reservoirs with examples of fluctuations in stored water volume. Symposium on the effects of storage on water quality. Water Research Centre, Medmenham.

Gorham, E. (1958). The physical limnology of northern Britain: an epitome of the Bathymetrical survey of the Scottish freshwater lochs, 1897–1909. *Limnol. Oceanogr.* 3, pp. 40–50.

Leopold, L. B., Wolman, M. G., and Miller, J. P. (1964). Fluvial processes in geomorphology. Freeman & Co., San Francisco and London.

Maitland, P. S. (1977). A coded checklist of animals occurring in fresh water in the British Isles. Institute of Terrestrial Ecology. Natural Environment Research Council, Edinburgh.

Mill, H. R. (1895). Bathymetrical Survey of the English Lakes. *Geogr. Journ.* 6, pp. 46–73 and 135–166.

Murray, J. and Pullar, L. (1910). Bathymetrical Survey of the Scottish Freshwater Lochs. Challenger Office, Edinburgh.

Strahler, A. N. (1957). Quantitative analysis of watershed geomorphology. *Trans. Amer. Geophys. Union* 38, pp. 913–920.

Water Resources Board and Scottish Development Department. (1974). The Surface Water Year Book, 1966–70. HMSO, London.

Printed in England for Her Majesty's Stationery Office by
Staples Printers St Albans Limited at The Priory Press